YOUR KNOWLEDGE HAS VALUE

- We will publish your bachelor's and
  master's thesis, essays and papers

- Your own eBook and book -
  sold worldwide in all relevant shops

- Earn money with each sale

Upload your text at www.GRIN.com
and publish for free

Peter Klapper

Adsorption equilibria of di(2-ethylhexyl)phosphoric acid at the water-dodecane-interface. Importance of zinc extraction

GRIN Verlag

Bibliografische Information der Deutschen Nationalbibliothek:

Die Deutsche Bibliothek verzeichnet diese Publikation in der Deutschen National-
bibliografie; detaillierte bibliografische Daten sind im Internet über http://dnb.d-
nb.de/ abrufbar.

Imprint:

Copyright © 2014 GRIN Verlag GmbH
Druck und Bindung: Books on Demand GmbH, Norderstedt Germany
ISBN: 978-3-656-75756-6

This book at GRIN:

http://www.grin.com/en/e-book/281462/adsorption-equilibria-of-di-2-ethylhexyl-
phosphoric-acid-at-the-water-dodecane-interface

Adsorption equilibria of di(2-ethylhexyl)phosphoric acid at the water-dodecane-interface: Part 2. Importance of zinc extraction

Keywords: Zinc extraction, cation exchanger, interfacial activity, adsorption equilibrium, pseudo-nonionic modelling, interfacial tension

Abstract

In this study the interfacial activity of the cation exchanger complex, which is fully loaded with zinc, is presented. The importance of the extraction equilibria on the equilibrium interfacial tension is explained by adding different electrolyte concentrations and by varying the concentrations of the cation exchanger di(2-ethylhexyl)phosphoric acid (D2EHPA).

In part 1 of this series, a modelling strategy of the equilibrium interfacial tension is validated for the two-phase system water-electrolyte/dodecane-D2EHPA in the case of the absence of metal salt extraction. This modelling strategy is applied to the extended system with zinc extraction. For this purpose, the previously used Gibbs-Duhem equation is extended by the influences of the two further components - the zinc complex without ligands and the zinc ion. The Langmuir isotherm of the multicomponent adsorption is used to describe the adsorption of the surfactants. The adsorption of the ionic derivate of the cation exchanger induces a counterion adsorption. This caused adsorption is formulated by the Stern isotherm. A simple model describes the considerable aggregation tendency of the D2EHPA complexes, which are fully loaded with zinc. This model is analogous to the model of micelle formation of the cation exchanger anion. All in all, the presented model reproduces acceptably the measured equilibrium interfacial tension.

1 Introduction

In liquid-liquid extraction the concentration profiles of the two different liquids merge continuously into each other the interface. The interface is not a two-dimensional field but a volume with a small thickness. If a local concentration maximum of a substance exists across the interface thickness, the substance is called a surfactant. Because of the resulting concentration elevation in the interface in comparison to the two liquids these substances determine the properties of the interface. The interfacial concentration describes the accumulation of a substance in the interface and it represents an averaging of the concentration profile across the interfacial thickness. At equilibrium the interfacial concentration is correlated with the activities of the surfactants in the two liquids by an adsorption isotherm. Since the adsorbed substances change the interfacial tension between the liquids, this material parameter is used to describe the

"

material accumulation in the interface. The interfacial tension is an essential material parameter, which determines the drop behaviour in disperse systems and influences the design and the equipment of technical extractors [1,2].

The modelling of the adsorption equilibria is especially important for formulating the non-equilibrium processes in the liquid-liquid extraction with interfacial active reactants or rather reactive surfactants. It is well known for the adsorption of surfactants that the accumulation of a substance can be done from the bulk phase into the interface with and without kinetic inhibition [3,4,5]. If the adsorption is immediately from the interface-near volume to the interface, the mass transfer into the interface is determined by the adsorption equilibrium. In the case of a kinetic inhibited transfer sorption kinetic relations are used for modelling. For the steady state of the adsorption these relations can be converted into the adsorption isotherms. That is why the sorption kinetics can be derived from the adsorption equilibria.

In the first part of this series [6], the adsorption equilibria of di(2-ethylhexyl)phosphoric acid (D2EHPA) at the water-dodecane-interface is presented under the condition of missing metal salt extraction. In this part the developed model is upgraded to the zinc extraction.

In the literature the misconception is often presented that the interfacial tension depends only on the cation exchanger concentration during the zinc extraction [7,8]. However, our own studies [9] show that the non-solvated zinc complex is a surfactant but not the zinc complexes, which are solvated with D2EHPA. Because of the solvation tendency of the zinc complexes, the zinc complexes are found only in the solvated form, if sufficient unloaded D2EHPA molecules exist in the organic solution. Under these conditions, the interfacial tension is determined by the unbound D2EHPA molecules in the form of monomers and their anions. The interfacial activity of the zinc complex without ligands can only be studied for an almost completely loaded cation exchanger [10].

2 Experimental

Before the tensiometric measuring starts, both phases were mixed at a constant volume ratio in a shaker for 24 hours. After this procedure the equilibrium state is reached between the two phases. In principle, systematic inaccuracies can result from the different volumes of the phases in the several tensiometric experiments. Because of the equilibrium state these methodical fault can be prevented.

The zinc was extracted from sulphate solutions. In addition to varying the initial concentrations of the cation exchanger and the zinc sulphate in the corresponding liquids, the extractions were performed with the addition of various quantities of sulphuric acid and sodium sulphate. The pH value was measured to reduce the number of mass balance equations.

The D2EHPA initial concentrations of the organic solutions were varied between 1 mmol/l and 1 mol/l. All electrolyte additives were diversified between 10^{-4} and 10^{-1} mol/l in the aqueous solution before starting the mixing of the liquids.

The measurement of the interfacial tension was carried out at 20°C on an aqueous pendant drop, which was hanging in a cuvette filled with dodecane. The steady value of the measured dynamic interfacial tension profile is the equilibrium interfacial tension. The drop shape analysis is used as a physical principle. In this method, the drop shape of a rotationally symmetrical drop, which is recognized as a silhouette, is theoretically adjusted by the Gauss-Laplace equation with the interfacial tension as a fitting parameter [11]. The principle experimental setup is illustrated in the first part of this series [6].

With the exception of the cation exchanger (D2EHPA, 95 weight-%, technical quality, Sigma) the used chemicals were checked for possible contaminations by tensiometric measurements and in the case of the diluent (n-dodecane, 99 weight-%, for synthesis, Merck) a preparation was necessary. The diluent was cleaned by repeated washing with deionized water. The other chemicals were used in analytical quality.

For the determination of the extractive equilibrium data 10 ml organic solution was mixed with 20 ml aqueous solution for 24 hours. After phase separation the pH value was measured in the aqueous phase. Following the zinc concentration was determined in the aqueous sample by inductively coupled plasma-atomic emission spectroscopy (Perkin Elmer ICP-OES PE Optima 3000). The zinc concentration of the organic phase was calculated by the concentration change in comparison to the initial value of the zinc concentration in the aqueous solution. The D2EHPA initial concentrations were varied between 10^{-3} and 1 mol/l for this test series, because the extracted quantities of zinc must be greater than the error of chemical analysis in the aqueous sample. All added electrolytes – zinc sulphate, sodium sulphate, sodium hydroxide, sulphuric acid – were varied according to the concentrations of the tensiometric measurements.

3 Modelling

Unlike the first part of the series [6] the adsorption of the zinc complex without ligands must be considered in addition to the immediate adsorption of the monomer and the cation exchanger anion. Also the consequential adsorption of the counterions must be supplemented by the influence of zinc ions in comparison with the system without zinc sulphate addition. Because of these extensions and neglecting any electrostatic influences of the electrochemical double layer it is obtained for the Gibbs adsorption equation:

$$d\gamma = -\Re T\left[\Gamma_{\overline{HR}}\frac{da_{\overline{HR}}}{a_{\overline{HR}}} + \Gamma_{H^+}\frac{da_{H^+}}{a_{H^+}} + \Gamma_{R^-}\frac{da_{R^-}}{a_{R^-}} + \Gamma_{Na^+}\frac{da_{Na^+}}{a_{Na^+}} + \Gamma_{\overline{ZnR_2}}\frac{da_{\overline{ZnR_2}}}{a_{\overline{ZnR_2}}} + \Gamma_{Zn^{2+}}\frac{da_{Zn^{2+}}}{a_{Zn^{2+}}}\right] \tag{1}$$

In this equation the Langmuir isotherms of the multicomponent adsorption Eq. (2) until Eq. (4) are introduced.

$$\Gamma_{\overline{HR}} = \Gamma_{\infty,\overline{HR}}\frac{K_{L,\overline{HR}}\,a_{\overline{HR}}}{1 + K_{L,\overline{HR}}\,a_{\overline{HR}} + K_{L,R^-}\,a_{R^-} + K_{L,\overline{ZnR_2}}\,a_{\overline{ZnR_2}}} \tag{2}$$

$$\Gamma_{R^-} = \Gamma_{\infty,R^-}\frac{K_{L,R^-}\,a_{R^-}}{1 + K_{L,\overline{HR}}\,a_{\overline{HR}} + K_{L,R^-}\,a_{R^-} + K_{L,\overline{ZnR_2}}\,a_{\overline{ZnR_2}}} \tag{3}$$

$$\Gamma_{\overline{ZnR_2}} = \Gamma_{\infty,\overline{ZnR_2}}\frac{K_{L,\overline{ZnR_2}}\,a_{\overline{ZnR_2}}}{1 + K_{L,\overline{HR}}\,a_{\overline{HR}} + K_{L,R^-}\,a_{R^-} + K_{L,\overline{ZnR_2}}\,a_{\overline{ZnR_2}}} \tag{4}$$

The adsorption of the counterions is a consequence of the adsorption of the cation exchanger anion and it is described by the Stern isotherms:

$$\Gamma_{H^+} = \frac{K_{S,H^+}\,a_{H^+}}{1 + K_{S,H^+}\,a_{H^+} + K_{S,Na^+}\,a_{Na^+} + K_{S,Zn^{2+}}\,a_{Zn^{2+}}}\Gamma_{R^-} \tag{5}$$

$$\Gamma_{Na^+} = \frac{K_{S,Na^+}\,a_{Na^+}}{1 + K_{S,H^+}\,a_{H^+} + K_{S,Na^+}\,a_{Na^+} + K_{S,Zn^{2+}}\,a_{Zn^{2+}}}\Gamma_{R^-} \tag{6}$$

$$\Gamma_{Zn^{2+}} = \frac{K_{S,Zn^{2+}}\,a_{Zn^{2+}}}{1 + K_{S,H^+}\,a_{H^+} + K_{S,Na^+}\,a_{Na^+} + K_{S,Zn^{2+}}\,a_{Zn^{2+}}}\Gamma_{R^-} \tag{7}$$

If these isotherms and the link between anion, proton and monomer, which reduces the variables on the independent activities [6],

$$\frac{da_{R^-}}{a_{R^-}} = \frac{da_{\overline{HR}}}{a_{\overline{HR}}} - \frac{da_{H^+}}{a_{H^+}} \tag{8}$$

are introduced into the Gibbs adsorption equation Eq. (1), the following relation results for the equilibrium interfacial tension after integration:

$$\gamma = \gamma_0 - \Re T \frac{\Gamma_{\infty,\overline{HR}} K_{L,\overline{HR}} a_{\overline{HR}} + \Gamma_{\infty,R^-} K_{L,R^-} a_{R^-}}{K_{L,\overline{HR}} a_{\overline{HR}} + K_{L,R^-} a_{R^-}} \ln\left[1 + \frac{K_{L,\overline{HR}} a_{\overline{HR}} + K_{L,R^-} a_{R^-}}{1 + K_{L,\overline{ZnR_2}} a_{\overline{ZnR_2}}}\right]$$

$$- \Re T \Gamma_{\infty,\overline{ZnR_2}} \ln\left[1 + \frac{K_{L,\overline{ZnR_2}} a_{\overline{ZnR_2}}}{1 + K_{L,\overline{HR}} a_{\overline{HR}} + K_{L,R^-} a_{R^-}}\right]$$

$$- \Re T \frac{\Gamma_{\infty,R^-} K_{L,R^-} a_{R^-} K_{S,H^+} a_{H^+}}{(1 + K_{S,Na^+} a_{Na^+} + K_{S,Zn^{2+}} a_{Zn^{2+}})(1 + K_{L,\overline{HR}} a_{\overline{HR}} + K_{L,\overline{ZnR_2}} a_{\overline{ZnR_2}}) - K_{L,R^-} a_{R^-} K_{S,H^+} a_{H^+}}$$

$$\cdot \ln \frac{1 + \dfrac{K_{L,R^-} a_{R^-}}{1 + K_{L,\overline{HR}} a_{\overline{HR}} + K_{L,\overline{ZnR_2}} a_{\overline{ZnR_2}}}}{1 + \dfrac{1 + K_{S,Na^+} a_{Na^+} + K_{S,Zn^{2+}} a_{Zn^{2+}}}{K_{S,H^+} a_{H^+}}}$$

$$- \Re T \frac{\Gamma_{\infty,R^-} K_{L,R^-} a_{R^-}}{1 + K_{L,\overline{HR}} a_{\overline{HR}} + K_{L,R^-} a_{R^-} + K_{L,\overline{ZnR_2}} a_{\overline{ZnR_2}}} \ln\left[1 + \frac{K_{S,Na^+} a_{Na^+}}{1 + K_{S,H^+} a_{H^+} + K_{S,Zn^{2+}} a_{Zn^{2+}}}\right]$$

$$- \Re T \frac{\Gamma_{\infty,R^-} K_{L,R^-} a_{R^-}}{1 + K_{L,\overline{HR}} a_{\overline{HR}} + K_{L,R^-} a_{R^-} + K_{L,\overline{ZnR_2}} a_{\overline{ZnR_2}}} \ln\left[1 + \frac{K_{S,Zn^{2+}} a_{Zn^{2+}}}{1 + K_{S,H^+} a_{H^+} + K_{S,Na^+} a_{Na^+}}\right] \qquad (9)$$

In deriving this relation, the reactive coupling between the monomer and the zinc complex is neglected. This simplification is possible because the zinc complex occurs only at high loadings, which are associated with very low monomer concentrations [10].

In the previous equations the activity of the zinc complex means the activity of the disaggregated complexes. The aggregation of the organic zinc complexes in the dodecane phase is formulated analogously to the micelle forming reaction of the D2EHPA anion in the aqueous phase [6]. The micelle formation of the D2EHP itself is irrelevant under the chosen conditions because of the zinc extraction from acid solutions.

$$n \cdot \overline{ZnR_2} \Leftrightarrow \left(\overline{ZnR_2}\right)_n \qquad (10)$$

On the basis of the mass action law of this reaction

$$K_{Ag,\overline{ZnR_2}} = \frac{a_{\left(\overline{ZnR_2}\right)_n}}{a_{\overline{ZnR_2}}^n} \qquad (11)$$

the fraction of the disaggregated zinc complex can be determined by the implicit calculation rule using the total activity of the zinc complex:

$$\frac{1 - \alpha_{\overline{ZnR_2}}}{\alpha_{\overline{ZnR_2}}^n} = K_{Ag,\overline{ZnR_2}} a_{\Sigma,\overline{ZnR_2}}^{n-1} \qquad (12)$$

This equation assumes that the aggregate formation is without influence on the chemical reaction equilibria. That is why the activity of the zinc complex aggregate is the same like the summation of the activity of the disaggregated complexes according to the aggregated number. From the fraction of the disaggregated zinc complexes the activity of the zinc complexes, which are important for the adsorption, is determined by the total activity of the zinc complexes:

$$a_{\overline{ZnR_2}} = \alpha_{\overline{ZnR_2}} \, a_{\overline{ZnR_2},\Sigma} \qquad (13)$$

The determination of the total activity of the zinc complex without ligands is based on the measured zinc extraction equilibria using the Wilson model for the formulation of the concentration dependence of the activity coefficients in dodecane. The various components with and without zinc loading are considered. Further differentiation of the unloaded cation exchanger is neglected and only the dimers are considered. Because of the pronounced tendency to dimerization of the cation exchanger in aliphatic diluents this approximation is permitted without a considerable loss of accuracy.

$$\ln \gamma_i = 1 - \ln\left(\sum_j x_j \cdot \Lambda_{ij} \right) - \sum_k \frac{x_k \cdot \Lambda_{ki}}{\sum_j x_j \cdot \Lambda_{kj}} \qquad \text{with} \qquad x_i = \frac{c_i}{\sum_j c_j} \qquad (14)$$

For calculating the equilibrium constants and the activity constants it is assumed simplistically, that the total amount of D2EHPA is constant. Thus the transfer of D2EHPA to the aqueous phase is neglected. Since the zinc extraction experiments were always performed in an acid regime and because of the low solubility of D2EHPA in acidic solution [10], this approximation does not lead to a considerable error. For analysing the interfacial tension as a result of the adsorption, however, these low amounts are important.

The interaction parameters are calculated by the equilibrium distribution of zinc and they are listed in Table 1.

	$C_{12}H_{26}$	$(HR)_2$	$ZnR_2(HR)_2$	$ZnR_2(HR)$	ZnR_2
$C_{12}H_{26}$	1	3,2173	0	0,101	0
$(HR)_2$	0,0381	1	0,0398	0	0
$ZnR_2(HR)_2$	0,0213	0	1	0	0
$ZnR_2(HR)$	0,1497	0	0	1	0
ZnR_2	0,1738	0	0	0	1

Table 1: Interaction parameter Λ of the Wilson model at 20°C

The three different types of zinc complexes are considered by the chemical equilibrium relations resulting from the mass action law [10]:

$$(HR)_2 + Zn^{2+} \Leftrightarrow ZnR_2 + 2H^+ \qquad \Rightarrow \qquad K_{Ex,0} = \frac{a_{\overline{ZnR_2},\Sigma} \cdot a_{H^+}^2}{a_{Zn^{2+}} \cdot a_{\overline{(HR)_2}}} \qquad (15)$$

$$1\tfrac{1}{2}(HR)_2 + Zn^{2+} \Leftrightarrow ZnR_2(HR) + 2H^+ \qquad \Rightarrow \qquad K_{Ex,1} = \frac{a_{\overline{ZnR_2(HR)}} \cdot a_{H^+}^2}{a_{Zn^{2+}} \cdot a_{\overline{(HR)_2}}^{1,5}} \tag{16}$$

$$2(HR)_2 + Zn^{2+} \Leftrightarrow ZnR_2(HR)_2 + 2H^+ \qquad \Rightarrow \qquad K_{Ex,2} = \frac{a_{\overline{ZnR_2(HR)_2}} \cdot a_{H^+}^2}{a_{Zn^{2+}} \cdot a_{\overline{(HR)_2}}^2} \tag{17}$$

The equilibrium constants of the various complexes are also derived from the equilibrium values of the zinc distribution. They are listed in Table 2.

$K_{Ex,0}$	$K_{Ex,1}$	$K_{Ex,2}$
$5{,}921 \cdot 10^{-3}$ mol/l	$1{,}869 \cdot 10^{-1}$ mol^{0,5}/l^{0,5}	$21{,}936$

Table 2: Chemical equilibrium constants of the zinc extraction at 20°C

After calculating the free cation exchanger concentration using the concentrations of the zinc bounded complexes and the initial concentration of D2EHPA the calculation of the activities of the monomer and cation exchanger anion is carried out. The electrolyte activities are estimated using the extended Debye-Hückel law on the basis of individual hydrated ionic radii [6].

$$\ln \gamma_i = -\frac{z_i^2 \Im^2 \cdot \kappa}{2\varepsilon_0 \varepsilon_{rel} \Re T N_A \left(1 + \kappa \cdot r_{i,hyd}\right)} \tag{18}$$

The used hydrated ionic radii are listed in Table 3.

H^+	Na^+	Zn^{2+}	OH^-	HSO_4^-	SO_4^{2-}	R^-
4,3 Å	3,5 Å	3,2 Å	4,8 Å	3,9 Å	3,7 Å	3,9 Å

Table 3: Used hydrated ionic radii r_{hyd}

4 Results

For the following simulation results, the calculated parameters of the equilibrium interfacial tension adjustment without zinc extraction [6] are adapted and only the additional constants of adsorption and aggregation are numerically fitted. These constants are listed in Table 4.

$\Gamma_{\infty,\overline{ZnR_2}}$	$K_{L,\overline{ZnR_2}}$	$K_{S,Zn^{2+}}$	$K_{Ag,\overline{ZnR_2}}$	n
$1{,}189 \cdot 10^{-4}$ mol/m^2	$5{,}029 \cdot 10^4$ l/mol	$1{,}649 \cdot 10^5$ l/mol	$21{,}936$ (mol/l)$^{1-h}$	10

Table 4: Fitted adsorption parameters of the zinc complex and the zinc ion at 20°C

The maximum interfacial concentration of the zinc complex is unusually high and significantly greater than that of the monomer or that of the anion [10]. At first this result is unexpected, because the reciprocal value is proportional to the minimum of interfacial space and the zinc complex is by far the largest molecule. However, it is known from the literature [12,13] that the loaded cation exchanger tends to the formation of polymers as a result of very strong intermolecular attractive interactions between the valence electrons of the oxygen of the organic phosphoric acid and the zinc atoms. Another effect of the strong intermolecular forces is the very high aggregation constant.

The importance of the Stern adsorption constant of the zinc ion is not insignificance, because the sum of squared errors of the fitted data increased by 30 % in the case of neglecting the zinc adsorption. This surprised, since higher concentrations of the zinc ions and the cation exchanger anions are not compatible because of the extraction constants. But since the concentration of the cation exchanger anion is very small in the investigated concentration range, the counterion adsorption of zinc is given a greater amount of impact than it is indeed the case. This is due to the insufficient description of the aggregate formation by only one aggregation form.

In the case of the only zinc sulphate addition (see Fig. 1) the calculated interfacial tension curves demonstrate the matching fitting of the measured curves by the presented model. Only for the high initial concentrations of zinc sulphate the model predicts two local extreme values, which are not found experimentally. Both the plateau region is recorded correctly and the interfacial tension of the other concentration ranges is reflected consistently. By supplementing the aggregation model with further polynary aggregated forms the remaining weaknesses could be resolved.

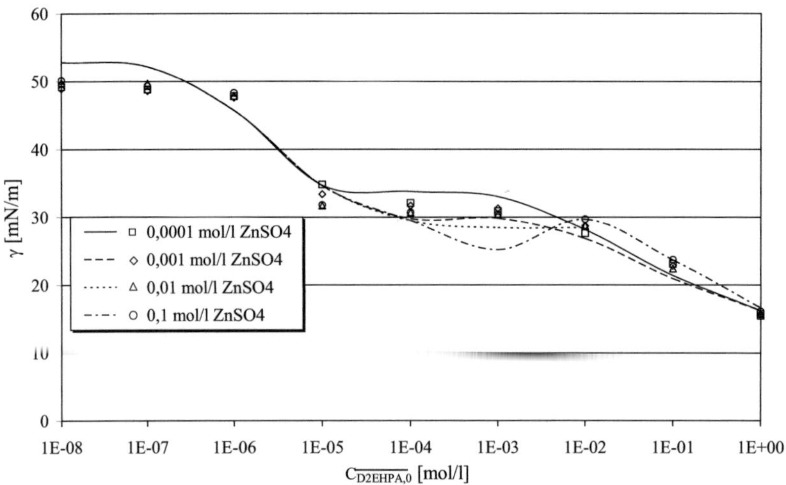

Figure 1: Equilibrium interfacial tension for various additions of zinc sulphate

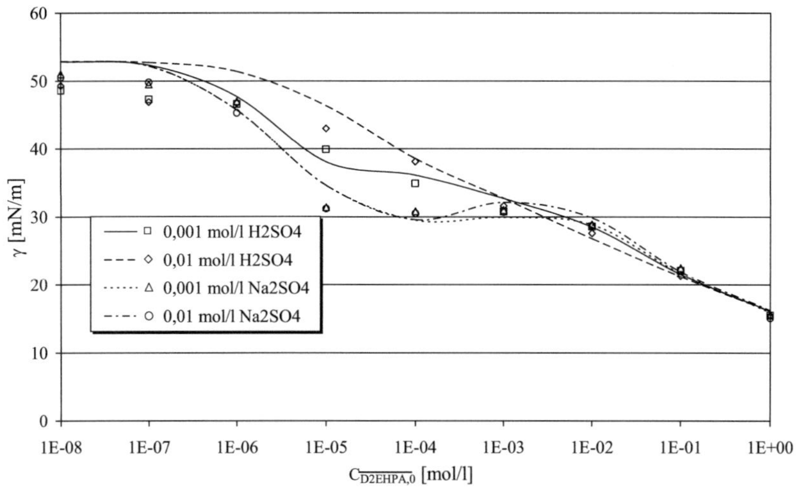

Figure 2: Equilibrium interfacial tension of a 0.01 mol/l zinc sulphate solution with various additional electrolytes

The performance of this model is also reflected in the further interfacial tension profiles. The interfacial tension can be reproduced consistently for the two different initial concentrations of zinc sulphate – 0.01 mol/l (see Fig. 2) and 0.001 mol/l (see Fig. 3) – and in the cases of adding different concentrations of sodium sulphate and sulphuric acid.

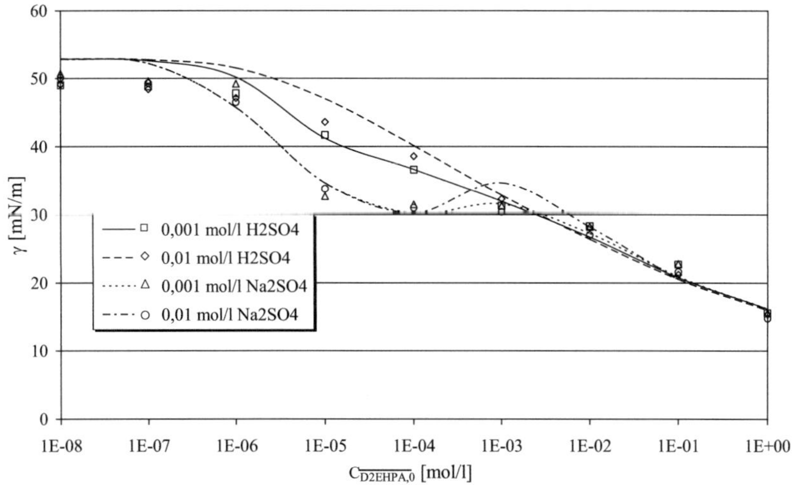

Figure 3: Equilibrium interfacial tension of a 0.001 mol/l zinc sulphate solution with various additional electrolytes

Qualitative differences between theory and experiment occur only with the addition of 0.01 mol/l sodium sulphate. In this case, the model predicts an oscillation of the interfacial tension at the measured plateau region. This simulated behaviour can also be a consequence of the incorrect modelling of the adsorption of ionic components in addition to the inadequate description of the aggregation of the zinc complexes.

The measured interfacial tension curves without additional acidification are of particular importance. Here, a strong decrease of the interfacial tension is observed in the range of the total D2EHPA concentrations between 1 μmol/l and 10 μmol/l. This decrease is much more intense than in the systems without zinc extraction. This effect and the fact, that the measured interfacial tension curves cannot be reflected consistently by the model if the zinc complex adsorption is neglected in modelling, confirm the strong interfacial activity of the zinc complex without ligands. The found plateau area and the measurements in the acidified material system verify the lack of interfacial activity of the zinc complexes with ligands. If the experimental conditions are chosen such that the zinc complexes are solvated with ligands, the interfacial tension can be described only by the adsorption of the monomer and the cation exchanger anion. It can be concluded from the lack of interfacial activity of the solvated zinc complexes and the strong interfacial activity of the zinc complex without ligands that the solvation takes place outside the interface after the desorption of the zinc complex without ligands.

5 Summary

A pseudo-nonionic modelling strategy is used to describe the equilibrium interfacial tension in the case of the zinc extraction with the cation exchanger D2EHPA. For this purpose a self-developed model, which is validated by a system without metal extraction, is supplemented. Besides the well-known interfacial activity of the monomer and the D2EHPA anion the zinc complex without ligands is considered as an additional interfacial active component. In the case of zinc extraction with D2EHPA these surfactants are regarded by using the Langmuir isotherms of the multicomponent adsorption in the formulation of the Gibbs-Duhem equation. Like in first part of this series the counterion adsorption of the electrolytes can be involved by the Stern isotherm, which describes the consequential adsorption because of the adsorption of the D2EHPA anion. Possible influences of the interfacial potential on the adsorption are neglected. The zinc complexes without ligands organize themselves as polynary aggregates. This pronounced tendency is described by a simple model. The presented model assumes that the aggregates have the same aggregation number and that the extraction equilibria are not influenced by the aggregate formation.

For the simulation of the equilibrium interfacial tension after the extraction from aqueous zinc sulphate solutions without the addition of other electrolytes, the measured interfacial tensions are reproduced well. Only at the high zinc sulphate concentration – 0.1 mol/l $ZnSO_4$ – strong deviations are found in the D2EHPA concentration range between 0.1 mmol/l and 10 mmol/l. These deviations are a consequence of the simple aggregation model. The influence of the addition of sodium sulphate or sulphuric acid to the systems of zinc extraction can be consistently described. Only at the low zinc sulphate concentration in combination with the high sodium sulphate concentration of the aqueous solution an inconsistency is found for a one mmolar D2EHPA concentration. In addition to the errors resulting from the simplified description of the aggregation, this discrepancy between the model and the measured data can also be caused by a faulty calculation of the adsorption of the cation exchanger anion and the derived counterion adsorption.

Nomenclature

Symbols

a	activity [mol/l]
c	concentration [mol/l]
I	ionic strength [mol/l]
K	adsorption and chemical equilibrium constant [depends]
n	stoichiometric factor of aggregation formation [-]
r	radius [m]
T	temperature [K]
x	mol fraction [-]
z	ionic charge [-]
γ	interfacial tension, activity coefficient [depends]
Γ	interfacial concentration [mol/m^2]
Δ	difference operator
κ	reciprocal Debye-Hückel length [1/m]
Λ	Wilson's interaction parameter [-]

Subscripts and superscripts

Ag	aggregation
Ex	extraction
hyd	hydrated
I, j, k	species
L	Langmuir

S	Stern
0	starting size
∞	infinite dilution, maximum

Constants

N_A	Avogadro's number ($6.0225 \cdot 10^{23}$ 1/mol)
ε_0	dielectric constant of vacuum ($8.8544 \cdot 10^{-12}$ C^2/Nm^2)
T	Faraday constant ($9.6487 \cdot 10^4$ C/mol)
Y	ideal gas constant (8.3143 J/(K·mol))

Appendix

After introducing the Stern isotherms Eq. (5) until Eq. (7) and together with the relation Eq. (8) the Gibbs adsorption equation Eq. (1) leads to:

$$d\gamma = -\Re T \left[\left(\Gamma_{\overline{HR}} + \Gamma_{R^-} \right) \frac{da_{\overline{HR}}}{a_{\overline{HR}}} + \Gamma_{Zn^{2+}} \frac{da_{Zn^{2+}}}{a_{Zn^{2+}}} \right.$$

$$- \frac{1 + K_{S,Na^+} a_{Na^+} + K_{S,Zn^{2+}} a_{Zn^{2+}}}{1 + K_{S,H^+} a_{H^+} + K_{S,Na^+} a_{Na^+} + K_{S,Zn^{2+}} a_{Zn^{2+}}} \Gamma_{R^-} \frac{da_{H^+}}{a_{H^+}}$$

$$+ \frac{K_{S,Na^+}}{1 + K_{S,H^+} a_{H^+} + K_{S,Na^+} a_{Na^+} + K_{S,Zn^{2+}} a_{Zn^{2+}}} \Gamma_{R^-} da_{Na^+}$$

$$\left. + \frac{K_{S,Zn^{2+}}}{1 + K_{S,H^+} a_{H^+} + K_{S,Na^+} a_{Na^+} + K_{S,Zn^{2+}} a_{Zn^{2+}}} \Gamma_{R^-} da_{Zn^{2+}} \right] \tag{A1}$$

This equation can be separated into one non-ionic part and three different ionic parts. After introducing the Langmuir isotherms Eq. (2) until Eq. (4) and the following algebraic integration the non-ionic part of the interfacial tension change can be described by:

$$\Delta\gamma_{\overline{HR},R^-,\overline{ZnR_2}} = -\Re T \left(\int \left(\Gamma_{\overline{HR}} + \Gamma_{R^-} \right) \frac{\partial a_{\overline{HR}}}{a_{\overline{HR}}} + \int \Gamma_{\overline{ZnR_2}} \frac{\partial a_{\overline{ZnR_2}}}{a_{\overline{ZnR_2}}} \right) + c_1$$

$$= -\Re T \left(\frac{\Gamma_{\infty,\overline{HR}} K_{L,\overline{HR}} a_{\overline{HR}} + \Gamma_{\infty,R^-} K_{L,R^-} a_{R^-}}{K_{L,\overline{HR}} a_{\overline{HR}} + K_{L,R^-} a_{R^-}} + \Gamma_{\infty,\overline{ZnR_2}} \right)$$

$$\cdot \ln\left(1 + K_{L,\overline{HR}} a_{\overline{HR}} + K_{L,R^-} a_{R^-} + K_{L,\overline{ZnR_2}} a_{\overline{ZnR_2}} \right) + c_1 \tag{A2}$$

The integration constant is calculated by the steady transition to the description without metal extraction, which is derived in the first part of the series [6].

The integration of the ionic parts of the interfacial tension resulting from the adsorption of the sodium ion and the zinc ion is carried out without additional transformations. The associated integration constants are defined by the criterion, that the part of the interfacial tension change resulting from the adsorption of a counterion is eliminated if this counterion does not exist in the aqueous solution.

$$\Delta\gamma_{Na^+} = -\Re T \int \frac{K_{S,Na^+}}{1 + K_{S,H^+} a_{H^+} + K_{S,Na^+} a_{Na^+} + K_{S,Zn^{2+}} a_{Zn^{2+}}} \Gamma_{R^-} \, da_{Na^+} + c_2$$

$$= -\Re T \Gamma_{R^-} \ln\left(1 + \frac{K_{S,Na^+} a_{Na^+}}{1 + K_{S,H^+} a_{H^+} + K_{S,Zn^{2+}} a_{Zn^{2+}}}\right)$$

(A3)

$$\Delta\gamma_{Zn^{2+}} = -\Re T \int \frac{K_{S,Zn^{2+}}}{1 + K_{S,H^+} a_{H^+} + K_{S,Na^+} a_{Na^+} + K_{S,Zn^{2+}} a_{Zn^{2+}}} \Gamma_{R^-} \, da_{Zn^{2+}} + c_3$$

$$= -\Re T \Gamma_{R^-} \ln\left(1 + \frac{K_{S,Zn^{2+}} a_{Zn^{2+}}}{1 + K_{S,H^+} a_{H^+} + K_{S,Na^+} a_{Na^+}}\right)$$

(A4)

$$\Delta\gamma_{H^+} = \Re T \int \frac{1 + K_{S,Na^+} a_{Na^+} + K_{S,Zn^{2+}} a_{Zn^{2+}}}{1 + K_{S,H^+} a_{H^+} + K_{S,Na^+} a_{Na^+} + K_{S,Zn^{2+}} a_{Zn^{2+}}} \Gamma_{R^-} \frac{da_{H^+}}{a_{H^+}} + c_4$$

(A5)

The following substitution rules are used for solving the remaining integration problem:

$$A^* = 1 + K_{L,\overline{HR}} a_{\overline{HR}} + K_{L,\overline{ZnR_2}} a_{\overline{ZnR_2}}$$

(A6a)

$$B^* = K_{L,R^-} \frac{K_a}{K_p} a_{\overline{HR}}$$

(A6b)

$$A = K_{S,H^+} \left(1 + K_{L,\overline{HR}} a_{\overline{HR}} + K_{L,\overline{ZnR_2}} a_{\overline{ZnR_2}}\right)$$

(A6c)

$$B = \left(1 + K_{S,Na^+} a_{Na^+} + K_{S,Zn^{2+}} a_{Zn^{2+}}\right)\left(1 + K_{L,\overline{HR}} a_{\overline{HR}} + K_{L,\overline{ZnR_2}} a_{\overline{ZnR_2}}\right) + K_{L,R^-} \frac{K_a}{K_p} a_{\overline{HR}} K_{S,H^+}$$

(A6d)

$$C = K_{L,R^-} \frac{K_a}{K_p} a_{\overline{HR}} \left(1 + K_{S,Na^+} a_{Na^+} + K_{S,Zn^{2+}} a_{Zn^{2+}}\right)$$

(A6e)

These rules are introduced in the tabulated solutions of indefinite integrals [14]:

$$\Delta\gamma_{H^+} = \Re T \Gamma_{\infty,R^-} \left(\int \frac{B^*}{a_{H^+}(A^* a_{H^+} + B^*)} \partial a_{H^+} - \int \frac{B^* K_{H^+}}{A a_{H^+}^2 + B a_{H^+} + C} \partial a_{H^+}\right) + c_4$$

$$= -\Re T \Gamma_{\infty,R^-} \left(\ln \frac{A^* a_{H^+} + B^*}{a_{H^+}} + \frac{B^* K_{H^+}}{\sqrt{B^2 - 4AC}} \ln \frac{2A a_{H^+} + B - \sqrt{B^2 - 4AC}}{2A a_{H^+} + B + \sqrt{B^2 - 4AC}}\right) + c_4$$

(A7)

For calculating the integration constants the same method is used as for the other two ions.

Literatur

[1] Bart, H.-J.; Maier, S.; Marr, R.; Weiß, S.: Apparateauswahl und Verwendung reaktionskinetischer Ansätze bei der Reaktivextraktion von Metallionen; Chemische Technik 45 (1993) 107-115

[2] Göttert, W.: Wissensbasierte Auswahl und Auslegung von Extraktoren; Shaker, 1993

[3] Chang, C.-H.; Franses, E. I.: *Adsorption dynamics of surfactants at the air/water interface: A critical review of mathematical models, data and mechanismus*; Colloids Surfaces A 100 (1995) 1-45

[4] Miller, R.; Aksenenko, E. V.; Liggieri, L.; Ravera, F.; Ferrari, M.; Fainerman, V. B.: *Effect of the reorientation of oxyethylated alcohol molecules within the surface layer on equilibrium and dynamic surface pressure*; Langmuir 15 (1999) 1328-1336

[5] Hsu, C.-T.; Chang, C.-H.; Lin, S.-Y.: *Comments on the adsorption isotherm and determination of adsorption kinetics*; Langmuir 13 (1997) 6204-6210

[6] Klapper, P.; Raatz, S.: *Adsorption equilibria of di(2-ethylhexyl)phosphoric acid at the water-dodecane-interface: Part 1. Effects of additional electrolytes*; GRIN Verlag, 2014

[7] Bart, H.-J.: *Reactive extraction*; Springer-Verlag, 2001

[8] Mörters, M.; Bart, H.-J.: *Extraction equilibria of zinc with bis(2-ethylhexyl)phosphoric acid*; J. Chem. Eng. Data 45 (2000) 82-85

[9] Raatz, S.; Klapper, P.: *Using interfacial tension measurements to analyze the mechanism of zinc extraction with D2EHPA*; Hydrometallurgy 134-135 (2013) 19-25

[10] Klapper, P.: *Tensiometrische Stofftransportuntersuchungen der Zinkextraktion mit dem Kationenaustauscher Di(2-ethylhexyl)phosphorsäure*; Thesis, TU Bergakademie Freiberg, 2010

[11] Song, B.; Springer, J.: *Determination of interfacial tension from the profile of a pendant drop using computer-aided image processing, 1. Theoretical*; J. Colloid Interface Sci. 184 (1996) 64-76

[12] Huang, T.-C.; Juang, R.-S.: *Extraction equilibrium of zinc from _ulphate media with bis(2-ethylhexyl)phosphoric acid*; Ind. Eng. Chem. Fundam. 25 (1986) 752-757

[13] Kunzmann, M.; Kolarik, Z.: *Extraction of zinc(II) with di(2-ethylhexyl)phosphoric acid from perchlorate and _ulphate media*; Solvent Extraction and Ion Exchange 10 (1992) 35-49

[14] Bronstein, I. N.; Semendjajew, K. A.: *Handbook of mathematics*; Springer, 2004